THEORY
of the
TOTAL UNIVERSE

MARVIN G. BACK

authorHOUSE®

AuthorHouse™
1663 Liberty Drive, Suite 200
Bloomington, IN 47403
www.authorhouse.com
Phone: 1-800-839-8640

First published by AuthorHouse 3/26/2009

ISBN: 978-1-4389-6271-9 (sc)

Printed in the United States of America
Bloomington, Indiana

This book is printed on acid-free paper.

DEDICATION

This book is dedicated to those individuals who provided me with inspiration and support:

My wife, Andriette, my daughter Maria and my granddaughter Sophia. Few know how they provide inspiration—a characteristic one may never knowingly intend but which manifests itself in both words and deeds that go well beyond cursory expression.

Also, to my friend Kim Melton, Ph.D. for her resourcefulness and constructive suggestions.

ABOUT THE COVER

The cycle of the Sub-Universe progresses from a dark and bleak origination within the Total-Universe (12 o'clock), to the Big Bang (2 o'clock) to greater and greater expansion then ultimately fades back into the Total-Universe. The stretched hands of the Infinity Clock unrelentingly mark the passage of time.

Light rays in the lower portion race out of the Sub-Universe and into the darkness of the Total Universe where they temporarily excite particles of Dark Matter (DM) and are heard from no more. There is no "light at the end of the tunnel," in a cosmic sense because the DM within the Total Universe continuously absorbs each photon—one at a time.

From the small sphere (representing the visible universe), within the yellow Sub-Universe, at the 4 o'clock position to the more massive patch at the 6 o'clock position represents the ability of humans to see farther and farther back into time. 8 o'clock remains in the future—just how far we do not know; however, given that our sun will swallow the Earth about five billion years into the future, if mankind survives until then, we may have extracted from our meager minds a capability to locate another world on which to continue our search.

ABOUT THE AUTHOR

Marvin G. Back was born in Sault Ste. Marie, Michigan, February 25, 1936. He is the youngest of three sons born to Albert W. and Fanny A. Back.

He holds an Associate in Science Degree from Kellogg Community College (Battle Creek, Michigan) and a Bachelors Degree in Business Management from Northwood University (Midland, Michigan). He is also a graduate of the Air War College, the National Defense University, the Army's Command and General Staff College and several branch qualification courses.

In 1963 he was commissioned in the Army Reserve with a total career spanning over 37 years during which time he attained the rank of Major General.

While serving with the Army, Back was awarded the Distinguished Service Medal, the Legion of Merit, the

Bronze Star Medal, the Meritorious Service Medal, the Army Commendation Medal and several others.

In business he served in managerial positions including a number at the executive level.

Back resides with his wife, Andriette, in Dahlonega, Georgia.

TABLE OF CONTENTS

INTRODUCTION

The "Theory of the Total-Universe" is my version of how Our-Universe came into being and which provides answers currently missing from fundamental cosmological theory. Questions like: What came before the singularity? Why did the universe expand as it did and still is? Will it continue to expand forever? Will it reverse course and end in a big crunch? Where is all the matter? Is there a fundamental matter form of which the universe is made (my personal question)? Will it expand for some finite time and then "go cold?" My theory answers these questions and provides a framework for cosmologists, astronomers and theoretical physicists to expand their thoughts beyond present limits.

This thesis is intentionally brief to ensure the fundamental premise is communicated without having to interpret extraneous data or complicated formulae. Note that the illustrations are designed to be representative, not to precisely describe size or shape.

Current, widely accepted, scientific theory purports that both our universe and time began with the big bang. However, this theory fails to spell out how the super-dense singularity came into existence. As defined, the singularity was the point from which everything in Our-Universe emanated and the moment it exploded (started to expand) was the beginning of time.

THEORY OF THE TOTAL UNIVERSE

While subscribing to the fundamental premise of the big bang, I, nevertheless, begin from a vastly different starting point: the time prior to and leading up to the big bang. In other words, the Total-Universe existed prior to the big bang and Our-Universe, I theorize, is a result of an event within the Total-Universe. Also note that for the purposes of my explanation, a theory is an idea or thought that requires further proof. I will outline some potential proofs later in the text.

At the outset, a trio of definitions may be helpful to ensure that the magnitude of the Total-Universe is placed in proper perspective:

Definition 1. The Total-Universe includes everything that exists.

Definition 2. Our-Universe is that area of expansion, within the Total-Universe, caused by the big bang.

From this point forward, Our-Universe will be replaced by the term Sub-Universe.

Definition 3. The Visible-Universe is that area of the Sub-Universe which humans can currently detect.

The beginning point for my theory is when the Total-Universe was near absolute zero, essentially stable and virtually totally dark (prior to the Big Bang). Within this infinite space-time exists a substance commonly referred to as dark matter (DM). I theorize that this substance was originally arranged as follows: somewhat like gas introduced into a void, DM tended to be distributed such that a relatively uniform homogeneity existed throughout the Total-Universe. The DM was in a very low state of activity because of the temperature and distances between particles (see illustration #1).

After the passage of time (for the sake of discussion I will assume a trillion, billion years--quite probably longer), and by sheer chance, particles of DM collided. This was accomplished without a significant release of energy and without the expulsion of any significant form of radiation. However, instantaneously a new entity was at work within the Total-Universe.

Keep in mind that DM (within the undisturbed Total-Universe) was extremely minute, had limited polarity, was very cold (near absolute zero) and consequently, in its normal state, reacted very slowly to the application of gravitational or other attractive forces. In their pre-collision state, DM particles may have been separated by minute or immense distances.

Nevertheless, when the DM particles formed a partnership they in effect initiated the beginning of our singularity (note that this event did not take place at any specific location within the Total-Universe and I do not define the size of a singularity). This new entity was attractive to other particles of DM; and due to this attraction and an almost imperceptible deflection in the universal fabric of space-time, a general migration of DM began to take place toward our singularity (Note that the term our singularity is used to differentiate it from any other singularities that may simultaneously or independently exist in the Total-Universe). This process continued over multi-billion year periods with increasing speed as the DM aggregation and resultant attraction continued to build until enough DM had accumulated at the singularity to cause an expansion. This expansion was minor indeed when viewed in the context of the Total-Universe albeit a tremendously creative explosion in human terms.

Imagine, if you will, the Total-Universe is a spheroid several trillion light years across (infinite comes to mind). Now, insert a grain of sand into the spheroid and call that the size of what I now refer to as a Sub- Universe. I do, in fact, theorize the Sub-Universe, though continuing to expand, is a finite structure.

A side comment may be helpful here to further comprehend the size of the Total-Universe. More than one singularity and sub-universe may have come into existence prior to ours or subsequent to ours and many may exist simultaneously. Locations, configurations and conditions within other singularities or sub-universes are unknown.

Given my definition that the Total-Universe, prior to the expansion of any singularity (see illustration #1), consisted only of DM, I am led to the conclusion that everything in the Sub-Universe is merely DM in a revised state except for the DM that remained in its original state. In contrast to the prevailing theory that DM was formed as a result of the big bang, I theorize, that DM was the causative factor.

Also, it is important to note that what we refer to as the Visible-Universe is generally quite small when considered in the context of our Sub-Universe and very, very minute when viewed in the context of the Total-Universe. In other words, what to us was a monumentally enormous event (the big bang), turned out to be barely a ripple in the space-time of the Total-Universe. What did happen prior to the big bang, however, was that areas of increased DM were formed near the singularity due to the attraction which, consequentially, caused areas of lower concentrations of DM at greater distances from the singularity. For the sake of this discussion I will refer to these as areas of high density dark matter (HDDM) and areas of low density dark matter (LDDM). Since the gathering process did not necessarily proceed in a uniform way, one portion of the singularity was nearer to an area of HDDM and another portion of the singularity, relatively speaking, was nearer to an area of LDDM. This would be true regardless of the size of the singularity. This unequal concentration of DM is critical to understanding the shape of the expansion area and velocity of matter subsequent to and surrounding the big bang since both shape and velocity are determined by this association.

The concentration of HDDM would be greater near an area where DM was more prevalent in the Total-Universe and migration toward our singularity would be more pronounced there than in areas at greater distances or lower concentrations. In some areas of the Total-Universe DM particles may have been near one another while in other areas immense distances apart. (see illustration #1).

Once the big bang took place and the expansion began, it encountered, in one direction, an area of HDDM which tended to slow (relatively speaking) the expansion in that direction. As a result, that area of the expansion began to mushroom out and move more laterally as well as longitudinally from the epicenter of the singularity (see illustration #2). Conversely, the area of expansion toward the LDDM proceeded much more rapidly and with less resistance. This also caused the boundary area of the expansion, in the direction of the HDDM, to remain closer to the initial point of expansion (the singularity) than the boundary area in the direction of the LDDM area (see illustration #2).

My hypothesis is that initially the Total-Universe was nearly uniformly dispersed with DM. Somewhere (not necessarily at or near any specific location) within the Total-Universe of DM our singularity existed in a relatively stable form from the standpoint that it was not in a state of explosive expansion. Over eons of time enough DM was attracted to our singularity to cause it, after reaching critical mass, to expand (explode) until it created, in my definition, a Sub-Universe of the Total-Universe.

Prior to the initiation of expansion DM was drawn toward our singularity which, in turn, caused the DM concentration to be reduced between the undisturbed areas of the Total-Universe and the Sub-Universe and become more dense near the singularity. At this time the DM, in effect, became the self-triggering device for expansion of the singularity. Additionally, there is an alternate theory regarding the triggering of the big bang. Please see the included addendum sheet for that explanation.

As the singularity expanded, it pushed most of the surrounding DM away from the epicenter where it caused a buildup of DM ahead of the expansion. This buildup then became a region of HDDM (at least in one direction) located beyond but adjacent to the area of the newly formed Sub-Universe. DM caught in the turmoil of the big bang was modified into a gaseous state and ultimately coalesced into matter as we know it today.

Some DM remained in the "near void" between the original location of the singularity and the exterior boundary of the expansion while most was forced outward. It cannot be assumed that our singularity existed at any particular location within the Total-Universe nor can expansion be assumed to have occurred in a near perfect spherical manner because of the varying density of the DM from the singularity outward.

Over time the DM forced outward from the singularity began to migrate back into the "near void" and to filter into and through the space-time I refer to as our Sub-Universe and also into the space-time of the Visible-Universe. After considerable time this would result in a buildup of DM in

the Sub-Universe until a state of near equilibrium returned-
-similar to undisturbed portions of the Total-Universe (see
illustration # 5). This would occur despite the accelerating
expansion of the Sub-Universe. The velocity at which the
general expansion continues to take place varies with the
concentration of DM in its path (DM that has migrated back
into the sub-universe) and its proximity to the boundary of
the Sub-Universe. Note, however, that except in the region
immediately adjacent to the initial spheroid of expansion
(a momentary state), the DM concentration becomes less
in every direction (for a finite distance), but not necessarily
uniformly. It therefore appears that the Sub-Universe's
expansion will increase in speed for an extremely long
period of time but not infinitely.

Note also that expansion is essentially in a specific direction
(away from the epicenter) and some things (various revised
forms of matter and energy) are traveling faster than others.
This is consistent with the theory that DM is less dense
(LDDM) in some directions compared to others where
HDDM is more dominant. The migration of DM back
into the Sub-Universe is also a contributing factor which
will help slow and ultimately stop the expansion. While not
addressed here in detail, it must be assumed that a gigantic
"hole" exists somewhere in the Total-Universe since all
matter was blown outward from the singularity.[1]

Let me specifically address the issue of the accelerating
expansion of the Sub-Universe. Since the mass of the Sub-
Universe would indicate it should be slowing down because
of the gravitational attraction between bodies within the

---

[1] This sentence was added following the original copyright.

Sub-Universe; why would it not only be expanding but doing so at an ever increasing rate? Simply put, there is something out there "pulling" on it with a greater attractive force than the Sub-Universe's internal gravitational attraction. What could it be? Dark matter is the answer and in a concentration millions, billions or trillions of times more dense and massive than our Sub-Universe.

After concluding that the expansion was and is now taking place in a non-uniform way, I then concluded that the resulting shape of the Sub-Universe would be in the form of a modified bell (see illustration # 2). Therefore, the Sub-Universe, after 10–20 billion years, has reached its current configuration. Given the fundamental premise that shape is determined by the concentration of DM (both HDDM and LDDM), this means the Sub-Universe will expand more toward the LDDM area and less toward the HDDM area. This would result in the bell shape of the Sub-Universe being extruded more and more toward an area of LDDM as time progresses (see illustration #3).

Next, I felt it necessary to approximate a location for the Visible-Universe within the Sub-Universe in order to explain the phenomena of all galaxies moving away from each other and the more distant galaxies moving at velocities greater than the nearer ones. My conclusion is that the Visible-Universe would have to be in an area of the Sub-Universe where expansion was occurring in multiple directions and where DM accumulation was relatively low.

I also concluded that everything in the Visible-Universe was moving in generally the same direction (outward along multiple axes) with some things moving faster, relatively speaking, than others due to the attractive force of the DM within the Total-Universe.

With these parameters set I concluded that the Visible-Universe was   closer to an area of LDDM versus HDDM and in a location where space-time curvature and the continuously increasing distances would prevent currently available detection methods from being able to obtain a clear picture (see illustration #4).

Once technology and equipment reach the required level of capability, what we should eventually be able to "see" is an area of predominantly DM which will appear to be a nearly completely dark and cold region (nothing visible and only extremely minute radiation).

There also exists the possibility that another singularity or sub-universe could form, or has formed, relatively close to our Sub-Universe and that a singularity/sub-universe may ultimately collide with our Sub-Universe with unknown results. If another singularity underwent a dramatic explosion, like the big bang, near our Sub-Universe, a spacequake, of heretofore unknown proportions, would most likely occur.

CONCLUSIONS

1. The Total-Universe existed prior to the big bang.

2. A singularity of unknown size formed from dark matter within the Total-Universe and expanded (exploded) to form a Sub-Universe.

3. As currently envisioned, our universe, as I define it, is a Sub-Universe of the Total-Universe.

4. The Visible-Universe is totally within the space-time of the Sub-Universe which is within the space-time of the Total-Universe.

5. Dark matter is the substance from which everything in our Sub-Universe originated (except for the dark matter that remained in its original state).

6. The amount of dark matter in the Sub-Universe increases in proportion to its density in the area adjacent to the Sub-Universe.

7. The concentration of dark matter within the Sub-Universe will temporarily be greater near its perimeter boundary.

8. The rate of migration of dark matter into the Sub-Universe (from the Total-Universe) is dependent upon the DM's energy state, the pressure exerted by the expansion, and the attraction of elements within the Sub-Universe.

9. The energy state of dark matter within the Sub-Universe is greater than that in the undisturbed Total-Universe.

10. At some time the Total-Universe will return to a state of near equilibrium rather than ending in an inward big crunch or continuing to expand infinitely. What I predict will happen is that the attractive forces of DM within the Total-Universe will continue to pull on the matter of the Sub-Universe until it (the Sub-Universe) is consumed with such force that all matter once again reverts to DM and a state of equilibrium returns to the Total-Universe.

11. The question of parallel universes (multi-dimensions) is clearly possible since within the Total-Universe many sub-universes may exist simultaneously.

12. A formula for unification of the natural forces, using the Total Universe theory, should now be within the grasp of scientific thought.

POTENTIAL PROOFS

1. Inference and vision are, of course, contributors and tools which can and should be used in any scientific discourse, along with a healthy portion of common sense and logical reasoning.

2. A constructively critical examination completed by qualified people possessing knowledge that could support or modify the theory.

3. Dark matter will have to be isolated and used to determine a time of origination. If found to be in excess of 20 billion years old (current age of the sub-universe is pegged at 13 to 15 billion years), this would immediately add credibility to the theory that time began prior to the big bang. In fact, the isolation of dark matter may be the key clue that would unequivocally prove this theory because by my definition dark matter was the only substance in its original state carried through from the Total-Universe to the Sub-Universe.

4. The development of much more efficient tools is needed (i.e. something more efficient, by multiples, than charge-coupled devices or orbiting telescopes). Then, many observations by devices that can examine space-time beyond that which can be seen by current technology must be made in varying directions to determine whether or not we are close enough to the boundary of the expansion to view an area of dark matter. I predict that the vast majority of dark matter currently lies beyond both the boundaries of the Visible-Universe and the Sub-Universe. Caution must be exercised, however, since observation of the "hole" caused by the singularity expansion may result in erroneous data.

5. A wide range of long-distance spectral and radiation examinations would also need to be conducted in order to eliminate the possibility of the existence of material, other than dark matter, beyond the boundary of the Sub-Universe. The existence of another Sub-Universe within close proximity to ours would, however, have to be considered.

6. A precise determination of the quantity of dark matter near a convenient celestial body would have to be made and compared with results taken from a later time. An increase in dark matter would support the theory that dark matter is continuing to migrate back into the Sub-Universe.

7. Develop a method to very precisely verify the temperature of the sub-universe/visible-universe and compare that temperature with a future reading (using

the exact same criteria). I predict that over an extended time period, due to the proximal influence of the "Total Universe" (which surrounds the sub-universe/visible-universe) and, which, in accordance with my theory, is very near absolute zero, the temperature, as measured above, will decline. Because of periodic massive matter explosions, temperatures may locally and temporarily increase but will ultimately decline to near absolute zero.

8.   Launch a number of ultra-fast moving, extended-distance probes with the sole purpose of locating the boundary between the Sub-Universe and the Total-Universe.   On-board sensors would be encoded to detect minute, in-transit, temperature differentials and to relay that information to a collection station. The same probe must be capable of measuring the Dark Matter (DM) concentration at specified intervals while in-transit. I predict that the DM concentration will increase and the temperature will decrease as the Total-Universe boundary is approached.   As temperatures and DM data are reported from the probes, a definite direction to the boundary may be ascertained through comparative readings between probes and times.   Any single probe that reliably transmits a consistent decrease in temperature or increase in DM would provide the most likely direction to the nearest boundary position.

9.   Over time, galaxies will, I predict, begin to cool down and ultimately fade and go cold as dark matter continues its process of absorbing particles which inexorably

results in a protracted cool-down wherein dark matter asserts its influence with its methodical absorption of energy.

10. Due to the density of DM, light would be deflected as it passes through a DM gravitational field. This would cause light to zig-zag through space in the submicro-picometer range.

11. The continued acceleration of the mass of the Sub-Universe, being due to the attractive forces of the DM in the Total-Universe, will continue until the mass of the Sub-Universe collides with the DM of the Total-Universe. At that time the mass will be absorbed and reconverted to DM.

12. (Refer to Addendum # 2) Measurement of a galaxial drift would prove the existence of an attractive force strong enough to affect all matter within the Sub-Universe and the visible universe.

13. To ensure a positive approach toward validation of my theory, current thoughts and concepts must be "put on the shelf" to avoid prejudgment based on preference for existing theory.

ADDITIONAL THOUGHTS ON COSMOLOGY

The Total-Universe is infinite to earthbound humans. Should humankind survive for another five billion earth-years, it shall not be able to prove that the Total-Universe is or is not infinite. They will, however, I'm convinced, prove something exists beyond our Sub-Universe given enough time and technological advancement.

Once dark matter (DM) has been isolated and identified, humans will be able to resolve the perplexing problem of a "Theory of Everything."

Herein lays the roadmap to Einstein's cosmological constant. I'm convinced it's there, we just have to look.

ADDENDUM

An alternate theory for initiation of singularity expansion follows: Consider another mass (other than our singularity) of DM begins to accumulate. As its attractive forces increase and distance between particles decrease, a mass of DM (size unspecified) began its journey toward the location of our singularity. As it approaches the super-dense singularity, its velocity dramatically increases to near light speed. And in accordance with Special Relativity, the mass begins to increase and when sufficiently large it exerts its own attractive force which causes the two bodies to move toward one another and collide. At the moment of collision the expansion began and the Sub-Universe was formed. Time, relative to the formation of the Sub-Universe, also began at that moment.

ADDENDUM #2
(30 OCTOBER 2007)

My theory states that the Total Universe surrounds the Sub-Universe which includes the visible universe. Galaxies are dispersing at an ever increasing rate toward some unknown destination (according to most scholars on the visible universe). I, however, predict that the galaxies are, in fact, heading toward the Total Universe and the closer they get the faster their rate of closure. One possible proof is available presently if a concerted effort is made to determine if visible galaxies are being attracted in a specific direction. This will not be a dramatic shift in location due to the velocity and magnitude of distances; but, nevertheless, I predict a detailed investigation would show a general drift of galaxies within the visible universe toward a common destination. What may appear to be a random dispersal of galaxies is, in fact, a general drift toward the Total Universe. Our galaxy's location, velocity and proximity to the Total Universe would be the determining factors regarding which direction this drift is taking place. Determination would be to first make a selection of a number of "base" galaxies from which to make drift assessment. Measurement results would be most easily seen the farther apart the "base" galaxies are located. As instrumentation becomes more accurate and with an ability to look more precisely at distant objects, the drift would be more easily detected. Drift would be toward the nearest region of the Total Universe. Some galaxies may already have transitioned back into the Total Universe (refer to the Theory of the Total Universe). If galaxies in one portion of the visible universe are found to be drifting in a direction different from others, somewhere between them would be the approximate centerline (not necessarily the center) of the Sub-Universe.

Caution must be taken to ensure the possibility that a sufficient time lapse has not occurred which would mean the drift has already been concluded and the galaxies are now completely in the grip of the Total Universe's gravity. In this case, any detected drift would be minimal.

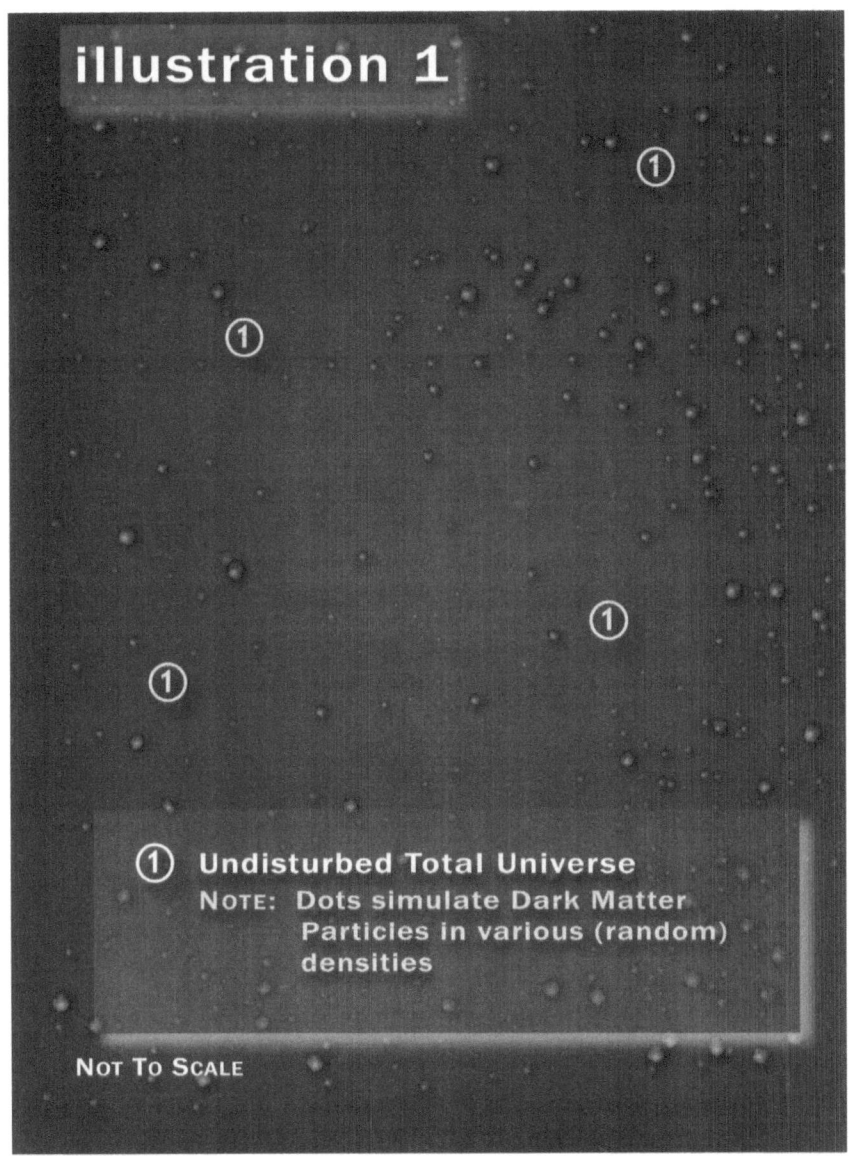

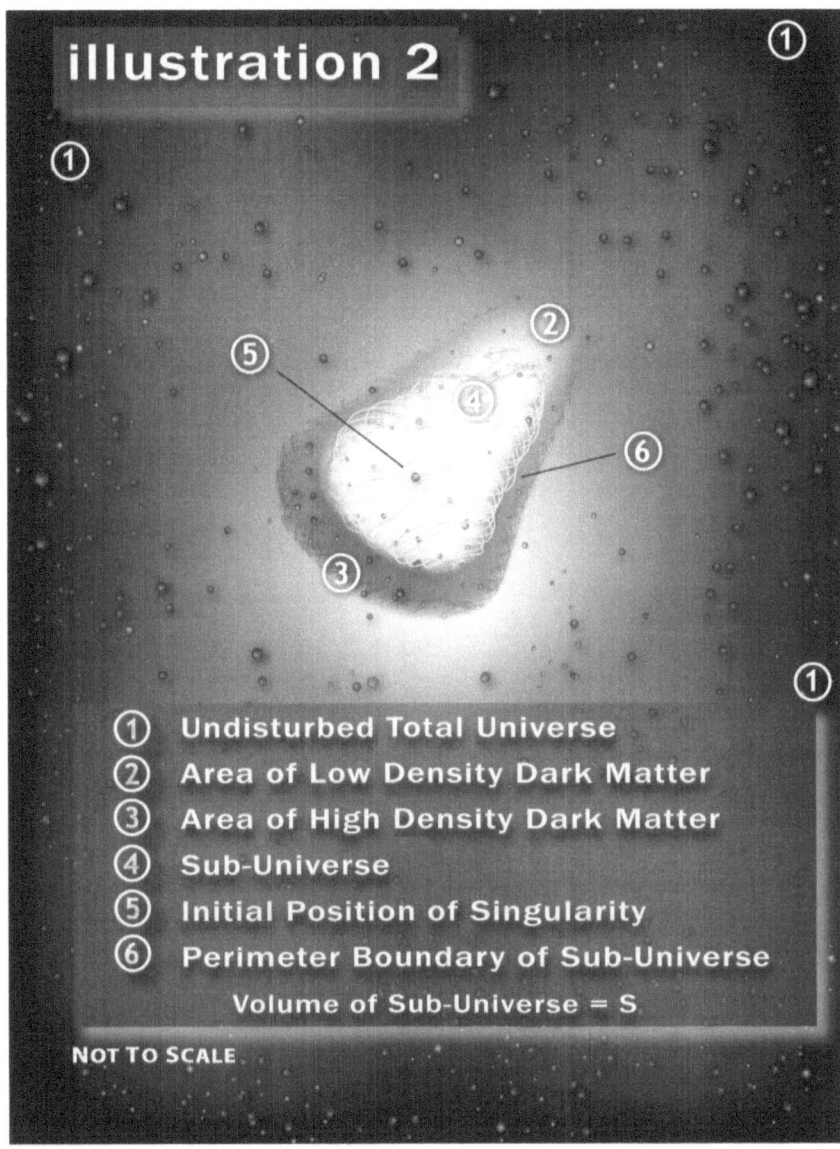

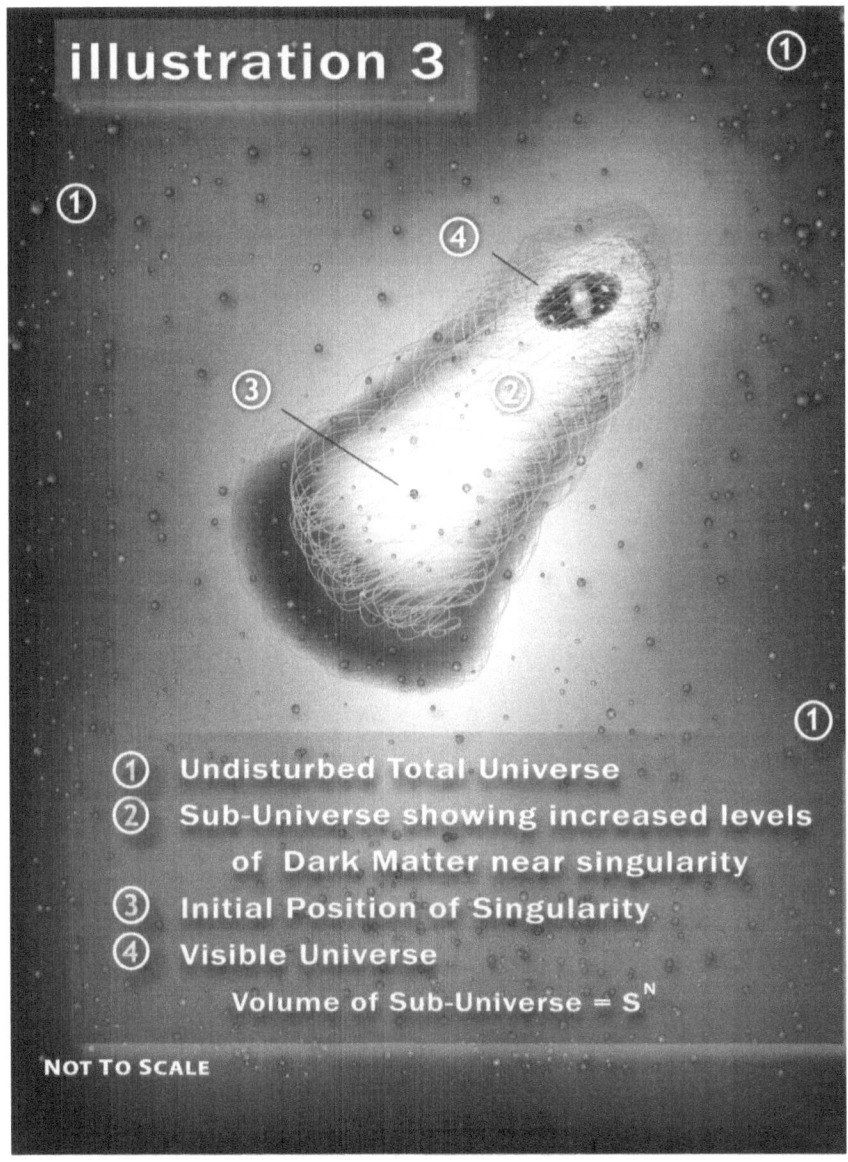

illustration 3

1. Undisturbed Total Universe
2. Sub-Universe showing increased levels of Dark Matter near singularity
3. Initial Position of Singularity
4. Visible Universe
   Volume of Sub-Universe $= S^N$

NOT TO SCALE

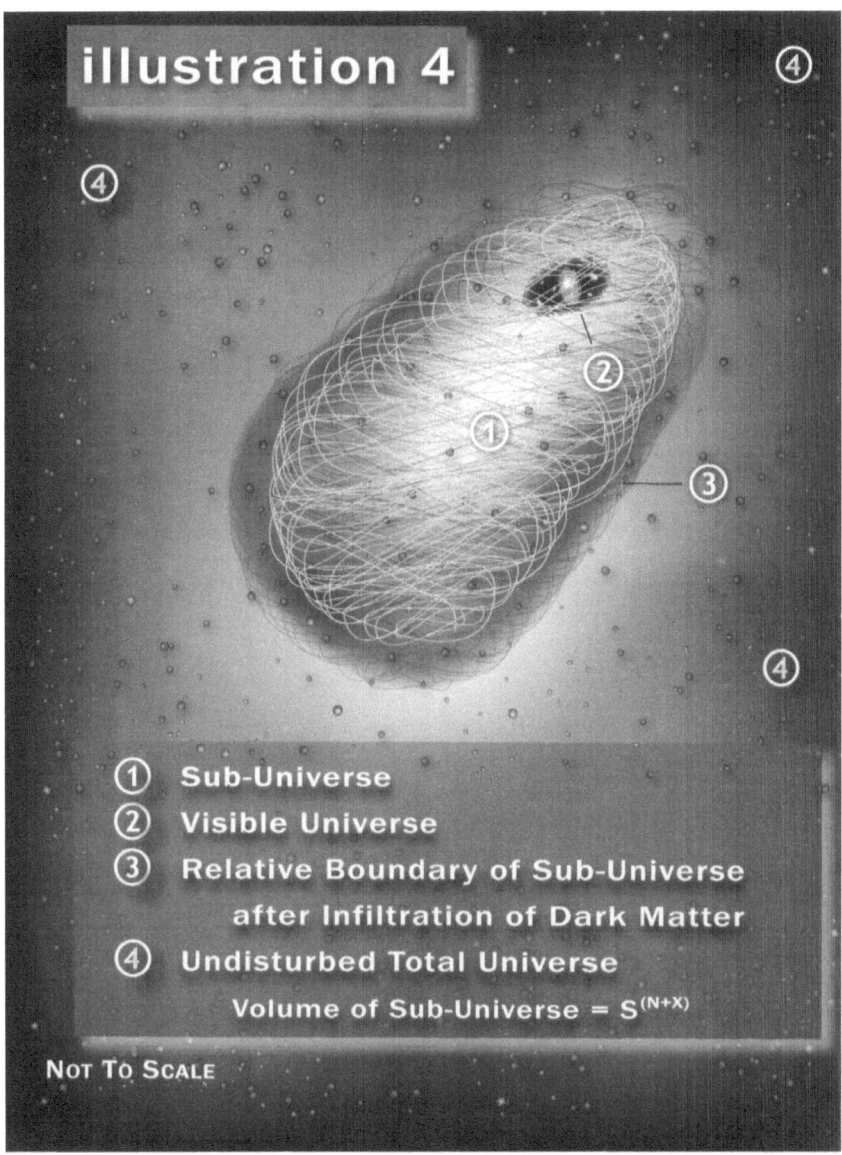

illustration 4

① Sub-Universe
② Visible Universe
③ Relative Boundary of Sub-Universe
   after Infiltration of Dark Matter
④ Undisturbed Total Universe
   Volume of Sub-Universe = $S^{(N+X)}$

NOT TO SCALE

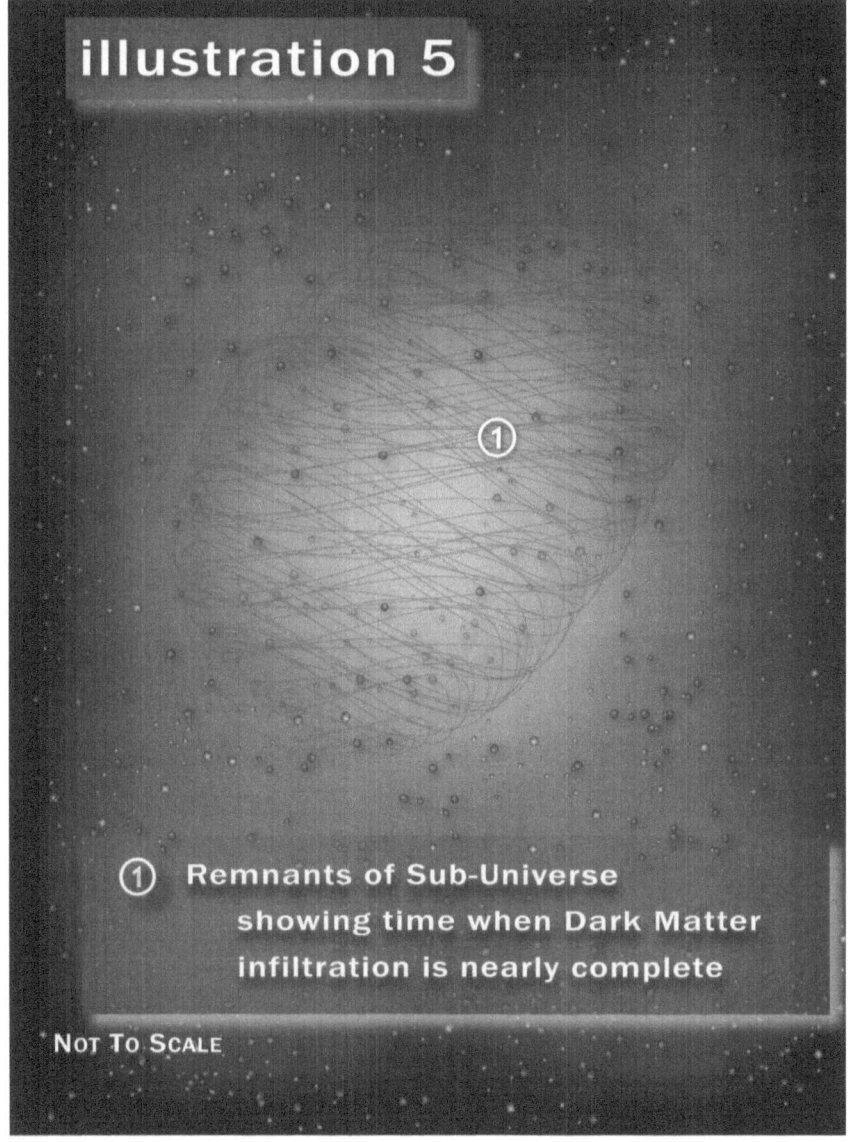

illustration 5

① Remnants of Sub-Universe
showing time when Dark Matter
infiltration is nearly complete

NOT TO SCALE

GLOSSARY

Absolute Zero — In 1848 William Thompson, 1st Baron Kelvin (Lord Kelvin) developed the Kelvin scale of absolute temperature measurement. Absolute zero being defined as the absence of all thermal energy. When referring to temperature on the Kelvin Scale, the term "degree" is not used. However, zero Kelvin (0 K) is approximately equal to -273.15 degrees Celsius or -459.67 degrees Fahrenheit

Big Bang — Sir Fred Hoyle, an English Astronomer, is credited with coining this term in 1949. It was, arguably, meant as a negative term as he referred to a theory, namely, The Big Bang, with which he disagreed. He, instead, believed in the Steady State Theory. The Big Bang Theory is an effort to explain what happened at the very beginning of "our universe." Note that in the text of this book, the author refers to "our universe" as a Sub-Universe of the Total Universe. This implies that neither the Steady State Theory nor the Big Bang Theory is totally correct.

Dark Matter — A Swiss Astrophysicist, Fritz Zwicky of the California Institute of Technology in 1933 inferred the existence of Dark Matter. At present it is thought that only about 4% of "Our Universe's" contents can be seen directly, 22% is believed to be in the form of Dark Matter, and the remaining 74% is Dark Energy. Here again, readers are referred to the book text wherein the author redefines "Our Universe" into "Sub-Universe."

Singularity — Sometimes referred to as the primeval atom and, according to current thought, is the point from which time, space, energy and matter originated. There is no definition (except in this book) as to how it (the Singularity) came into existence. Why it came into existence is a total mystery except in the beliefs of theologians.

Spacetime — Coined by Herman Minkowski in 1908, this term was later used by Albert Einstein when developing his Theory of Special Relativity. In 1926, the 13th edition of the Encyclopedia Britannica contained an article by Albert Einstein titled "Spacetime." This is a shortened version of the term spacetime continuum. Significant because it adds a fourth dimension to thoughts, which in the past, considered only three dimensions.

NOTES

NOTES

NOTES